The Jungle

Words by Carroll R. Norden, Ph.D.
Professor of Zoology
University of Wisconsin—Milwaukee

Raintree Childrens Books
Milwaukee • Toronto • Melbourne • London

Copyright © 1978, Macdonald-Raintree, Inc.

Library of Congress Number: 77-27590

4   5   6   7   8   9   0   82   81   80

Printed and bound in the United States of America.

Library of Congress Cataloging in Publication Data

Norden, Carroll R.
　The jungle.

　(Read about)
　Bibliography: p.
　Includes index.
　SUMMARY: Describes the characteristics and location
of different types of jungles and the people, plants,
and animals that inhabit them.
　1.　Jungles—Juvenile literature.　[1.　Jungles]
I.　Title.
QH541.5.J8M67　　　574.909'5'2　　　77-27590
ISBN 0-8393-0078-6 lib. bdg.

The Jungle

The jungle is a hot, damp place. It is a rain forest with a thick growth of trees and plants. The amount of rainfall and the temperature make it different from any other kind of forest.

In the jungle it rains nearly every day. The temperature stays about the same all the time. There are no real seasons.

Because of the rain and warm weather, plants grow very quickly. Most trees are evergreens. This means they have leaves all year round.

Leaves, flowers, and fruit are easy for plant-eating animals to get all year long. Other animals are meat eaters. They eat insects and other animals.

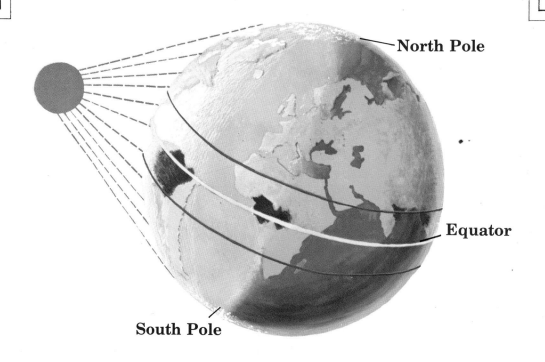

North Pole

Equator

South Pole

Jungles are found in the hottest parts of the world, near the equator. The equator is an imaginary line which circles the middle of the earth.

The sun's rays warm the earth. The shorter the distance the rays must travel to the earth, the warmer they are. The sun's rays have to travel a shorter distance to places near the equator. That makes places near the equator very hot.

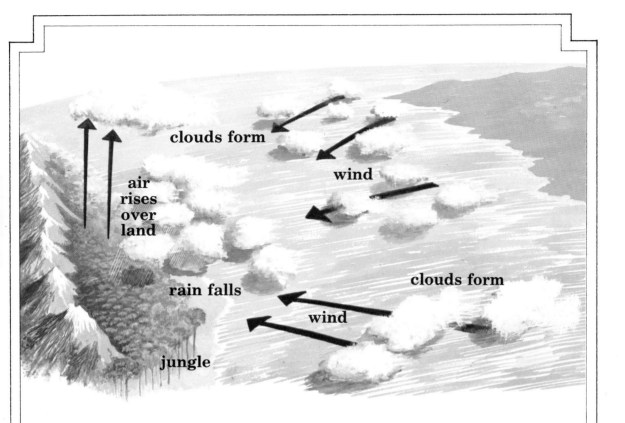

clouds form

air
rises
over
land

wind

rain falls

clouds form

wind

jungle

Besides being very hot, jungles are also very wet. Hot winds pick up fog and mist over the sea. When the winds reach the land, they rise and cool. Some of the fog and mist turn into tiny water drops. As the drops come together they form clouds. The drops become heavier than the air around them. Then the drops fall as rain.

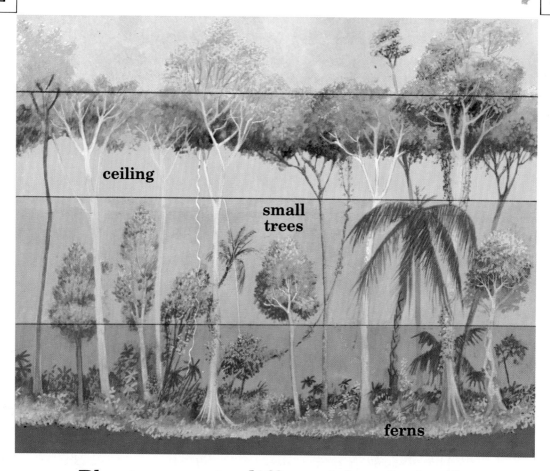

ceiling

small trees

ferns

Plants grow to different levels in the jungle. The tallest trees form a ceiling that keeps out the sunlight. Because little sunlight gets through the ceiling, the other levels are made up of plants that grow in the shade. The plants nearest the ground must grow in almost complete shade.

There are also different levels of animal life. Many jungle animals, such as birds, find their food high in the trees. They rarely come down to the ground. Some small animals live in the trees. Monkeys climb to the highest levels to get their food. Big animals, like elephants, tigers, and rhinoceroses, live on the jungle floor.

epiphyte   liana   strangler fig

Many large jungle trees have very wide roots. They are called buttress roots. These roots grow up the tree trunk. They support the trunks.

Epiphytes, another kind of plant often seen in jungles, grow along tree trunks and branches to reach the sunlight.

Thick, woody vines called lianas are also found in jungles. They hang in loops between the trees.

buttress root   fungi   fern

rafflesia

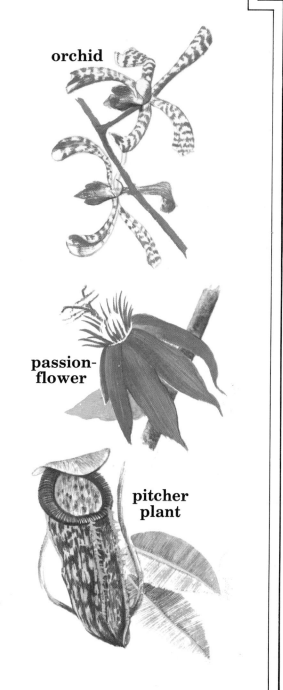

orchid

passion-
flower

pitcher
plant

One of the most interesting jungle plants is the rafflesia. It has the biggest flower in the world. Its flowers may be as large as 3 feet (about 1 meter) across.

Other brightly colored plants are the orchid and the passion-flower. They are often found high up on trees in the bright sunlight.

potto   hornbill   gorilla   guinea fowl   bush baby

mandrill

pangolin   okapi

Jungles are similar in many ways. But each jungle has its own kinds of plants and animals. Plants and animals found in an African jungle may not be found in a South American or Asian jungle.

These animals are often found in an African jungle.

chimpanzee

parrot

tree frog

dwarf antelope

giant forest hog

elephant

Many different kinds of monkeys and apes live in the African jungle. Apes such as chimpanzees and gorillas live mainly on the ground.

Many animals that live on the ground move by running, hopping, and leaping. Animals that live mostly in the trees may also climb, leap, and run. But they also may fly and glide.

There are small groups of Pygmies
living in the African jungle. They may live
in one place for about a month and then
move to another.

Pygmies build huts by covering a frame
of sticks with leaves. They use bows
and arrows to hunt. They fish with
nets and spears and gather fruits and
berries to eat.

macaw

harpy
eagle

marmoset

tamarin

boa

sloth

iguana

guinea
pig

The largest jungle in the world is in South America. Many of the plants and animals here are different from those in any other jungle.

Monkeys, sloths, and macaws are some of the animals that live in the South American jungle. The largest animals in the jungle are jaguars and tapirs. There are no large animals like apes, elephants, or rhinoceroses.

The longest river in South America runs through the South American jungle. It is called the Amazon River.

Tapirs and capybaras are two of the animals that live near the water. The tapir is one of the largest animals found in the jungle. The capybara is the largest rodent in the world. It looks like a giant guinea pig and weighs up to 100 pounds (about 45 kilograms).

anaconda

tapir

capybara

caiman

stork

Villages have been built along the Amazon River. Many were built by Amazon Indians. They build their houses on stilts. This helps keep them dry and safe when the river rises.

Amazon Indians gather berries, fruits, and nuts for food. They also fish and hunt animals. Their fishing boats are made from hollowed-out tree trunks.

flying
squirrel

proboscis
monkey

loris

rhinoceros

Asian elephant

Elephants and rhinoceroses are among
the largest mammals on earth. They are
found in the jungles of Asia.

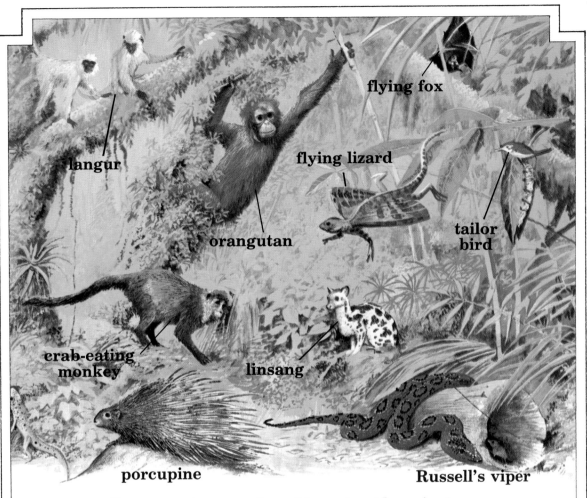

flying fox

langur

flying lizard

orangutan

tailor bird

crab-eating monkey

linsang

porcupine

Russell's viper

Orangutans also live in the Asian jungle. They spend most of their time in trees and are very clumsy on the ground.

They eat fruit, bark, and birds' eggs. Orangutans sleep in nests in trees. Each night they build a new nest.

Scientists often go to the jungles. They study the plants, insects, and animals. Sometimes they collect samples to take with them.

At night scientists may camp in the jungle. They build shelters from the wood and leaves. Often they sleep under mosquito nets. This protects them from insect bites.

Because the jungle air is almost always warm, it is a good place for the animals pictured on this page to live. All these animals are cold-blooded. That means their bodies are the temperature of the air or water around them.

These animals usually live in or near water. But in the jungle it is so wet that they can live in trees or on the wet floor of the jungle.

viper

flying lizard

croaking tree frog

Millions of insects live in the jungle. There are several thousand different kinds.

Ants are an important part of jungle life. Some insects live in nests made by ants. Some animals, like the anteater, feed on ants. Leaf-cutter ants bite out small pieces of leaves to take back to their nest.

Some mosquitoes are dangerous because their bite can spread diseases. Yellow fever and malaria are carried by mosquitoes. The largest spider in the world is the poisonous bird-eating spider. It lives in the jungle.

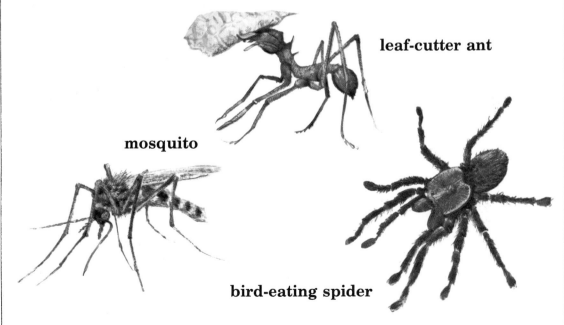

leaf-cutter ant

mosquito

bird-eating spider

Some jungle animals are nocturnal. This means they sleep during the day and hunt for food after dark. Many nocturnal animals have large eyes that help them see in the dark.

horned owl

margay

white bat

paca

jaguar

armadillo

Other jungle animals also have special parts of their bodies that help them live in the jungle.

The flying frog has wide webbed feet. They help it glide from tree to tree.

Some monkeys use their long, strong tails like an extra arm. It helps them to climb and swing from branch to branch. Some even hang by their tails.

flying frog

spider monkey

The hummingbird feeds on flowers. It has a long beak that lets it reach the nectar deep in the flower.

hummingbird

tiger

Some animals have different colors and patterns. This helps them to hide from their enemies. It also helps them get near other animals they are hunting without being seen.

gaboon viper

A tiger's stripes help it hide in tall grass. The color of some snakes is like that of the leaves they hide in. The see-through wings of the clearwing butterfly make it very hard to see when it is resting on a leaf.

clearwing butterfly

rubber trees　　　　　　　　　　**cup**

　　Many things that grow in jungles can be used by people.

　　　Rubber trees are found in jungles. Most rubber comes from Asia and South America.

　　　Rubber is made from a milky liquid called latex. A worker called a tapper cuts, or taps, the bark of the rubber tree. The latex flows into a cup. The latex is taken to a factory. There it is made into rubber products.

The cacao tree grows well in jungles. Beans in the cacao pods are used to make cocoa and chocolate.

The ripe cacao pods are cut from the trees. They are split open, and the beans are taken out. The beans are then cleaned and dried. They are sent to factories. There they go through many steps before they are made into cocoa and chocolate.

cacao beans

cacao pod

Other jungle trees are used as wood. Teak, mahogany, and ebony are woods that come from jungle trees. They are often used to make furniture and boats.

These trees have wide roots. To cut the trees down, men build platforms to reach the smaller part of the trunk.

In some jungles, trees are being cut down and burned. Roads are being built as the jungle is cleared.

The new roads
make it easier
for people to travel
in the jungle.

bananas

Some people clear the land and grow crops in the jungle.

Bananas grow in large bunches on thick stalks. There are about 100 different kinds of bananas.

oil palms

The fruit of the oil palm is also collected. It is crushed and used to make cooking oil.

pineapple

Pineapples are also grown by jungle farmers. They were first grown in South America. When different explorers came to South America, they took pineapples with them to other places. People in the West Indies used pineapples for food and to make wine. Other people used pineapples to make medicine. Today pineapples grow best in Hawaii, far from their first home.

The Metric System

In the United States, things are measured in inches, pounds, quarts, and so on. Most countries of the world use centimeters, kilograms, and liters for these things. The United States uses the American system to measure things. Most other countries use the metric system. By 1985, the United States will be using the metric system, too.

In some books, you will see two systems of measurement. For example, you might see a sentence like this: "That bicycle wheel is 27 inches (69 centimeters) across." When all countries have changed to the metric system, inches will not be used any more. But until then, you may sometimes have to change measurements from one system to the other. The chart on the next page will help you.

All you have to do is multiply the unit of measurement in Column 1 by the number in Column 2. That gives you the unit in Column 3.

Suppose you want to change 5 inches to centimeters. First, find inches in Column 1. Next, multiply 5 times 2.54. You get 12.7. So, 5 inches is 12.7 centimeters.

Column 1	Column 2	Column 3
THIS UNIT OF MEASUREMENT	TIMES THIS NUMBER	GIVES THIS UNIT OF MEASUREMENT
inches	2.54	centimeters
feet	30.	centimeters
feet	.3	meters
yards	.9	meters
miles	1.6	kilometers
ounces	28.	grams
pounds	.45	kilograms
fluid ounces	.03	liters
pints	.47	liters
quarts	.95	liters
gallons	3.8	liters
centimeters	.4	inches
meters	1.1	yards
kilometers	.6	miles
grams	.035	ounces
kilograms	2.2	pounds
liters	33.8	fluid ounces
liters	2.1	pints
liters	1.06	quarts
liters	.26	gallons

Where to Read About
the Jungle

gorilla (gə ril′ ə) *p. 13*
hummingbird (hum′ ing burd′) *p. 25*
jaguar (jag′ wär) *p. 15*
leaf-cutter ant (lēf′ kut′ ər   ant) *p. 23*
liana (lē än′ ə) *p. 10*
macaw (mə kô′) *p. 15*
monkey (mung′ kē) *pp. 13, 15, 25*
mosquitoe (məs kē′ tō) *p. 23*
oil palm (oil   päm) *p. 32*
orangutan (ô rang′ oo tan′) *p. 19*
orchid (ôr′ kid) *p. 11*
passionflower (pash′ ən flou′ ər) *p. 11*
pineapple (pīn′ ap′ əl) *p. 33*
Pygmy (pig′ mē) *p. 14*
rafflesia (rə flē′ zhə) *p. 11*
rain (rān) *pp. 4, 5, 7*
rhinoceros (rī nos′ ər əs) *p. 18*
rubber (rub′ ər) *p. 27*
sloth (slôth) *p. 15*
South American jungle (south   ə mer′ i kən
    jung′ gəl) *pp. 15-17*
tapir (tā′ pər) *pp. 15, 16*
temperature (tem′ pər ə chər) *pp. 4, 6*
tiger (tī′ gər) *p. 26*

Pronunciation Key

a	a as in **cat, bad**
ā	a as in **able,** ai as in **train,** ay as in **play**
ä	a as in **father, car,** o as in **cot**
e	e as in **bend, yet**
ē	e as in **me,** ee as in **feel,** ea as in **beat,** ie as in **piece,** y as in **heavy**
i	i as in **in, pig,** e as in **pocket**
ī	i as in **ice, time,** ie as in **tie,** y as in **my**
o	o as in **top,** a as in **watch**
ō	o as in **old,** oa as in **goat,** ow as in **slow,** oe as in **toe**
ô	o as in **cloth,** au as in **caught,** aw as in **paw,** a as in **all**
oo	oo as in **good,** u as in **put**
ō̄o	oo as in **tool,** ue as in **blue**
oi	oi as in **oil,** oy as in **toy**
ou	ou as in **out,** ow as in **plow**
u	u as in **up, gun,** o as in **other**
ur	ur as in **fur,** er as in **person,** ir as in **bird,** or as in **work**
yo͞o	u as in **use,** ew as in **few**
ə	a as in **again,** e as in **broken,** i as in **pencil,** o as in **attention,** u as in **surprise**
ch	ch as in **such**
ng	ng as in **sing**
sh	sh as in **shell, wish**
th	th as in **three, bath**
<u>th</u>	th as in **that, together**

GLOSSARY

These words are defined the way they are used in this book

amount (ə mount′) all of something added
together

ant (ant) a small insect

anteater (ant′ ē′ tər) a mammal that feeds
mostly or only on ants

ape (āp) a large, tailless monkey

beak (bēk) the hard part of a bird's mouth

berry (ber′ ē) the small, fleshy fruit of
a plant which can usually be eaten

body (bod′ ē) the whole of a person,
animal, or plant

bright (brīt) shining; giving off a lot
of light

built (bilt) made by putting a number of
things together

bunch (bunch) many things fastened or
growing together

butterfly (but′ ər flī′) an insect with a
thin body and four colored wings that
flies in the daytime

buttress root (but′ ris ro͞ot) the wide, thickened root of a tree that helps hold the tree up

cacao (kə ka′ ō) a South American evergreen tree that produces beans used to make chocolate and cocoa

ceiling (sē′ ling) the topmost level of something that forms a covering

chocolate (chô′ kə lit) a kind of food or drink made from roasted, ground up cacao beans

clear (klēr) to remove anything that is in the way

cloud (kloud) a mass of tiny drops of water or bits of ice that floats high in the sky

clumsy (klum′ zē) not graceful; awkward

cocoa (kō′ kō) the brown powder made from the dried beans of the cacao tree

cold-blooded (kōld′ blud′ id) having a body temperature that changes with the surrounding air or water temperature

collect (kə lekt′) to bring several things together

complete (kəm plēt') total; all of something

crop (krop) plants that are grown to be used for something helpful such as food

crush (krush) to press very hard

damp (damp) somewhat wet; not dry

disease (di zēz') illness or sickness

ebony (eb' ə nē) a hard, black wood that comes from some jungle trees

epiphyte (ep' ə fīt') a plant that grows on other plants and gets its food and water from the air and rain

equator (i kwā' tər) the imaginary line which circles the middle of the earth

evergreen (ev' ər grēn') a tree which has green leaves or needles all year round

explorer (eks plôr' ər) someone who travels to or examines unknown places to learn what they are like

extra (eks' trə) more than usually found

factory (fak' tər ē) a building or group of buildings where things are made

flow (flō) to move along in a steady stream

fog (fôg) a cloud made up of tiny bits of

water that is close to the ground

form (fôrm) to make or take shape

frame (frām) a support or border for an object

furniture (fur′ ni chər) certain moveable
objects, such as tables, chairs, beds, or
desks, used to make a place ready to live in

gather (ga<u>th</u>′ ər) to bring several things
together

glide (glīd) to float on air

growth (grōth) something that grows or
has grown

guinea pig (gin′ ē pig) a little, almost
tailless rodent with short ears and legs

hut (hut) a small building

imaginary (i maj′ ə ner′ ē) something that
is not real; make-believe

Indian (in′ dē ən) a member of one of the
tribes living in North and South America

insect (in′ sekt) a small animal with a
hard outer covering and without a backbone,
such as a fly or ant, and usually with
six legs and two or four wings

interest (in′ tər ist) to take and keep
the attention of

jungle (jung′ gəl) a hot, wet forest covered
  with many trees, plants, and vines
kilogram (kil′ ə gram′) a unit of measure
  equal to about two pounds and three ounces
latex (lā′ teks′) a milky fluid that is made
  by some plants such as the rubber tree
level (lev′ əl) to be even or at the same point
liquid (lik′ wid) a freely moving form of
  matter that is not a solid or a gas
mahogany (mə hog′ ə nē) a strong, hard
  wood that is yellowish to reddish brown
  in color and comes from some jungle trees
main (mān) the most important or the biggest
malaria (mə ler′ ē ə) a disease of warm
  places that is spread by mosquitoes
mammal (mam′ əl) a warm-blooded animal
  with a backbone and often with a growth
  of hair or fur
medicine (med′ ə sin) a drug or other
  material which is used to cure sickness
  or relieve pain
meter (mē′ tər) a measure of length equal
  to about 39 inches
milky (mil′ kē) like milk in color and thickness

million (mil′ yən) the number 1,000,000

mist (mist) small bits of water that float or fall in the air near the ground

nectar (nek′ tər) a sweet-tasting liquid formed inside a flower

net (net) loose material made of threads or cords knotted together so that there are many holes, used to catch a fish, insect, or other animal

nocturnal (nok turn′ əl) to move around and be lively at night

nut (nut) the dry shell-covered fruit of some plants

oil (oil) a greasy substance that does not mix with water

pattern (pat′ ərn) a way of putting together lines, colors, or shapes to make a design

platform (plat′ fôrm′) an even, flat surface higher than the surface around it

pod (pod) the part of some plants that holds the seeds

poisonous (poi′ zə nəs) able to cause sickness or death by poison

product (prod′ əkt) something that is made or created

rainfall (rān′ fôl′) all the rain that falls on a place during a certain time

rare (rer) not often taking place

ray (rā) a beam of light

ripe (rīp) fully grown; ready to be eaten

rise (rīz) to get up

rodent (rōd′ ənt) a small animal, such as a mouse or beaver, that has sharp front teeth used for cutting away at something little by little

root (rōot) the part of a plant that grows underground

rubber (rub′ ər) a stretchy substance that is made from the milky liquid of some plants

sample (sam′ pəl) part of something that shows what the larger whole or group is like

scientist (sī′ ən tist) someone who has studied a great deal about a branch of science

season (sē′ zən) one of the four times of the year — spring, summer, fall, and winter

shade (shād) a place where there is little or no sunlight

similar (sim′ ə lər) to be alike or much the same

snake (snāk) an animal with a long body usually covered with scales and having no legs, arms, or wings

spear (spēr) a throwing weapon with a sharp, pointed head attached to a long, thin handle

spider (spī′ dər) a small, wingless animal with four pairs of legs and a body divided into two parts that spins a web to trap insects for its food

split (split) to break into two or more parts

stalk (stôk) the main part of a plant from which leaves, flowers, and fruits grow

stilt (stilt) a long piece of wood or metal used to hold something above the ground or water

stripe (strīp) a long, narrow band

sunlight (sun′ līt′) the light that comes from the sun

support (sə pôrt′) to help keep in a certain place; hold up

teak (tēk) a hard yellowish brown wood that comes from some jungle trees

temperature (tem′ pər ə chər) the amount of heat or coldness

thousand (thou′ zənd) the number 1,000

travel (trav′ əl) to move from one place to another place

vine (vīn) a long, thin-stemmed plant that grows along the ground or up around the trunk of trees

webbed (webd) having the toes of a foot joined by skin growing between them

weigh (wā) to be or have a certain heaviness

wine (wīn) the drink made from the juice of some fruits, such as pineapples or grapes

woody (wood′ ē) to be all or partly made up of wood

yellow fever (yel′ ō fē′ vər) an easily spread disease of warm places carried by mosquitoes

Bibliography

Batten, Mary. *The Tropical Forest: Ants, Animals and Plants.* New York: Thomas Y. Crowell Co., 1973.

Johnson, Sylvia A. *Animals of the Tropical Forests.* Minneapolis, Minn.: Lerner Publications Co., 1976.

Ross, Wilda. *The Rain Forest – What Lives There.* New York: Coward, McCann and Geoghegan, Inc., 1977.

Silverberg, Robert. *World of the Rain Forest.* New York: Hawthorn Books, Inc., 1967.

Wellman, Alice W. *Africa's Animals: Creatures of a Struggling Land.* New York: G. P. Putnam's Sons, 1974.